FLORA OF TROPICAL EAST AFRICA

CORNACEAE

B. Verdcourt
(East African Herbarium)

Trees, shrubs or rarely perennial herbs. Leaves simple, alternate or opposite, entire or serrate ; stipules usually absent. Flowers hermaphrodite or dioecious, in racemes, panicles, umbels, clusters of cymules or compact heads. Calyx-tube adnate to the ovary ; sepals 4–5 or absent. Petals 4–5 or rarely absent, imbricate or valvate. Stamens 4–5, alternating with the petals. Ovary inferior, usually 2-celled but often 1–4 (–10)-celled ; ovules pendulous, one per loculus. Fruit a drupe or berry ; embryo small in abundant endosperm.

Only one genus with a single species occurs in East Africa. *Dendrobenthamia capitata* (Wall. ex Roxb.) Hutch. is cultivated in Kenya as an ornamental.

AFROCRANIA

(Harms) Hutch. in Ann. Bot., n.s., 6 : 89 (1942)

Cornus L. subgenus *Afrocrania* Harms in E. & P.Pf. III. 8 : 266 (1898)

Trees with opposite petiolate entire leaves. Flowers tetramerous, dioecious ; ♂ flowers in cymules which are arranged subumbellately in a dense terminal inflorescence ; ♀ flowers in terminal umbels ; both types of inflorescence have an involucre of four subherbaceous bracts which entirely include the very young inflorescence forming an apiculate bud. Peduncles thickened at the apex. Petals valvate. Ovary 2-locular. Fruit a drupe crowned by the remains of calyx and style ; stone 2-celled.

A. volkensii (*Harms*) *Hutch.* in Ann. Bot., n.s., 6 : 90 (1942) ; T.T.C.L. : 173 (1949) ; I.T.U., ed. 2 : 101 (1952). Type : Tanganyika, Kilimanjaro, Kilema, *Volkens* 1821 (B, holo.†)

Tree, usually ± 18 m. tall but attaining 24 m., bole almost straight ; bark grey or black, granular. Young stems finely adpressed pubescent later glabrous. Leaves light green, elliptic, acuminate at the apex, ± cuneate at the base, 5–17·5 cm. long and 2·5–6·3 cm. wide, finely and somewhat silvery adpressed pubescent, glabrescent with age above, often also with tufts of hairs in the nerve-axils below ; lateral nerves 4–5, arcuate-ascending, impressed above, prominent below ; petiole up to 2 cm. long. Flowers 20–100, greenish-yellow or cream, in dense inflorescences; peduncles of both ♂ and ♀ inflorescences 2·5–5 cm. long, silvery pubescent ; involucre-bracts ovate or orbicular, acuminate, 7–8 mm. long and 6–7 mm. broad, densely adpressed pubescent outside, soon falling. Cymules of ♂ inflorescences with pedicels 7–12 mm. long and secondary peduncles up to 4 mm. long ;

FIG. 1. *AFROCRANIA VOLKENSII*—**1,** leafy shoot with flowers of ♂ inflorescence in bud, × 1 ; **2,** part of ♂ inflorescence, × 3 ; **3,** ♂ flower, × 6 ; **4,** petal of ♂ flower, × 6 ; **5,** ♂ flower with petals removed, × 6 ; **6,** young ♀ inflorescence enclosed by involucre, × 1·5 ; **7,** ♀ inflorescence, × 3 ; **8,** ♀ flower, × 6 ; **9,** petal of ♀ flower, × 6 ; **10,** ♀ flower cut open longitudinally to show placentation, × 6 ; **11,** fruit, × 2 ; **12,** diagrammatic longitudinal section of fruit, × 2. 1, from *Stolz* 1979; 2–5, from *Ross* 924 ; 6, from *Eggeling* 1050 ; 7–10, from *Procter* 170 ; 11, 12, from *Dyson* (in spirit)

sepals broadly deltoid 0·5 mm. long ; petals 2 mm. long and 0·8 mm. wide, sparsely pubescent. Pedicels of ♀ flowers 2–9 mm. long, adpressed silvery pubescent ; calyx-tube (ovary) 3 mm. long, pubescent, minutely 4-toothed at apex ; petals glabrous or sparsely pubescent, ± 3 mm. long and 1 mm. broad. Fruit purplish-black when ripe, ellipsoidal, 13 mm. long and 6·5 mm. broad (in dried specimens 9–11 × 5 mm., and irregularly ridged) minutely adpressed pubescent. Fig. 1.

UGANDA. Ruwenzori, Bujuku Valley, Aug. 1931 (fl. & fr.), *Fishlock & Hancock* 173 ; Kigezi District : NW. face of Mt. Muhavura, *Eggeling* 1050 ; Mbale District : Elgon, Benet, Jan. 1936 (fr.), *Eggeling* 2458 !
KENYA. Nakuru District : Kikuyu Escarpment, Dec. 1934 (fl.), *Wimbush* 846 ! ; E. Aberdare Mts., Kerita, May 1930 (fl.), *Dale* 402 ! ; Mt. Kenya, *Hutchins in F.D.* 406 !
TANGANYIKA. Mbulu District : Oldeani, Nou Forest Reserve, Dec. 1950 (fl.), *Bond* 9 ! ; Lushoto District : Shume-Magamba Forest Reserve, June 1951 (fr.), *Eggeling* 6165 ! ; Rungwe District : Mt. Rungwe, Mar. 1932 (fl.), *St. Clair-Thompson* 853 !
DISTR. **U**2, 3 ; **K**2–4, 6 ; **T**2, 3, 6, 7 ; Belgian Congo, Nyasaland and Southern Rhodesia.
HAB. Upland rain-forest, associated with *Arundinaria, Ocotea* etc., often riparian ; tolerant of cold and heavy rainfall ; 1200–3000 m.

SYN. *Cornus volkensii* Harms in P.O.A. C : 301 (1895) and in E. & P.Pf. III. 8 : 266 (1898) ; Wangerin in E.J. 38, beibl. 86 : 6, 25, etc. (1906) and in E.P. IV. 229 : 76, fig. 19 (1910) ; V. E. 3 (2) : 835, fig. 338 (1921)

INDEX TO CORNACEAE